RISING STARS
Mathematics

Year **2**

Practice Book **C**

Author: Paul Broadbent

ISBN: 978-1-78339-815-7
Text, design and layout © Rising Stars UK Ltd 2016

First published in 2016 by
Rising Stars UK Ltd, part of Hodder Education,
An Hachette UK Company
Carmelite House
50 Victoria Embankment
London EC4Y 0DZ
www.risingstars-uk.com

Author: Paul Broadbent
Programme consultants: Cherri Moseley, Caroline Clissold, Paul Broadbent
Publishers: Fiona Lazenby and Alexandra Riley
Editorial: Jan Fisher, Aidan Gill
Project manager: Sue Walton

Series and character design: Steve Evans
Text design: Words & Pictures
Illustrations by Steve Evans

Cover design: Steve Evans and Words & Pictures

Printed by Liberduplex, Barcelona
A catalogue record for this title is available from the British Library.

Contents

10a Exploring faces

1 Complete this chart. Write each letter in the correct box.

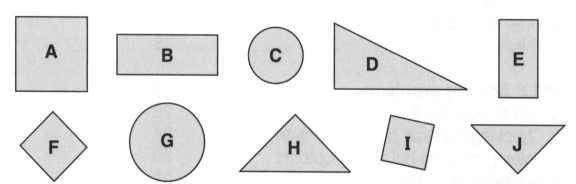

square	triangle	rectangle	circle

2 Name the shape of each shaded face.

a
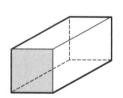

c
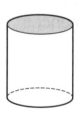

e
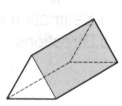

b
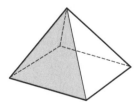

d
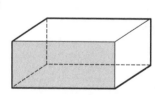

f
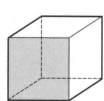

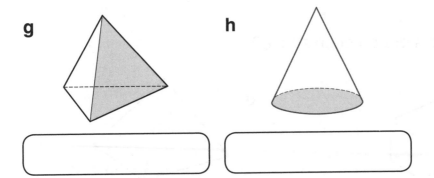

g [ ]

h [ ]

 3 Sort these shapes. Write each letter in the correct box.

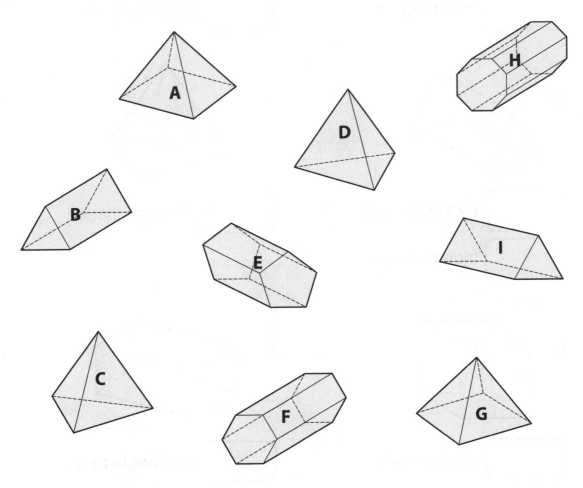

Prisms	Pyramids

5

4 How many faces are on each shape?

a
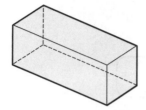

[ ] rectangle faces

[ ] square faces

d
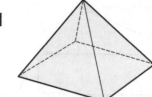

[ ] triangle faces

[ ] square faces

b
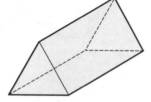

[ ] rectangle faces

[ ] triangle faces

e
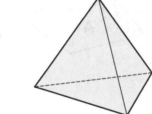

[ ] triangle faces

c
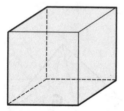

[ ] square faces

f
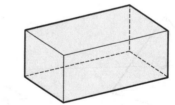

[ ] rectangle faces

Draw lines to show where each shape goes on this Venn diagram.

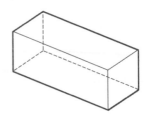
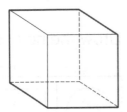
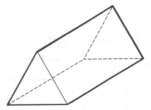

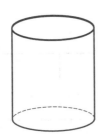
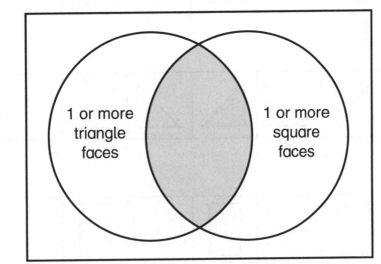

1 or more
triangle
faces

1 or more
square
faces

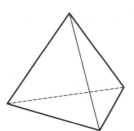
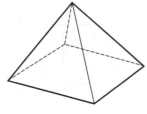
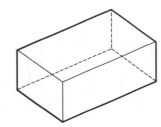
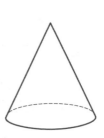

10b **Patterns and shapes**

1 Draw the next 3 shapes in each pattern.

a

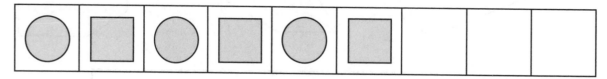

b

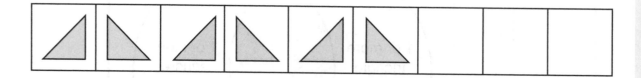

c

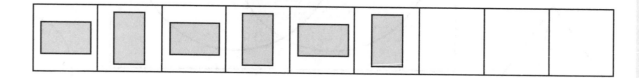

d

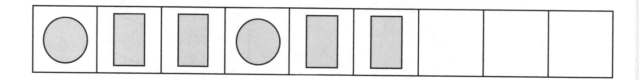

e

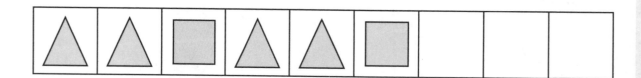

f

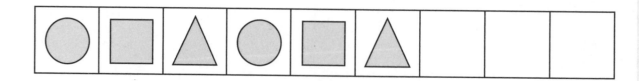

2 This L-shape can be used to make a tiling pattern.

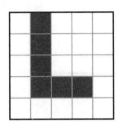

Use the L-shape to make your own tiling pattern here.

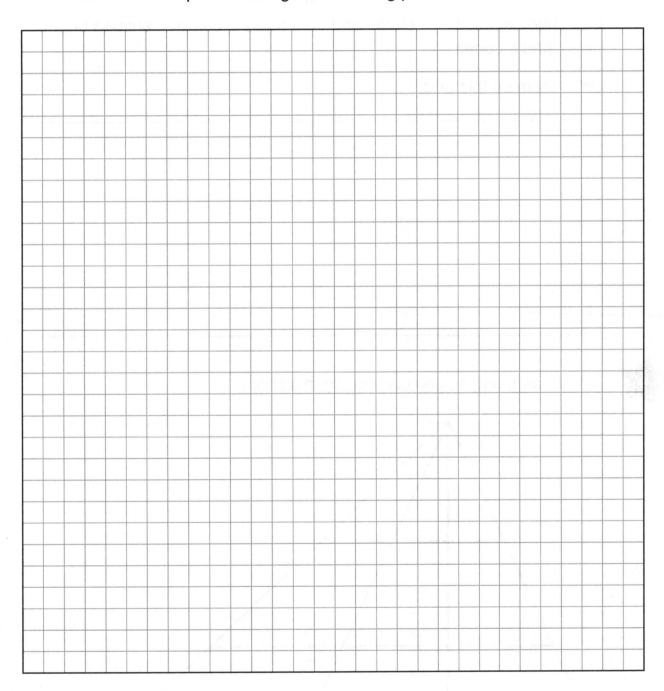

YOU WILL NEED:
• interlocking cubes

These 5 cubes are joined together to make this shape.

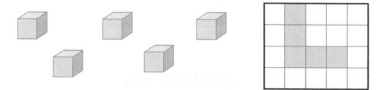

Use cubes to make the shape and then make 3 different shapes of your own using 5 cubes.

Draw them on this grid.

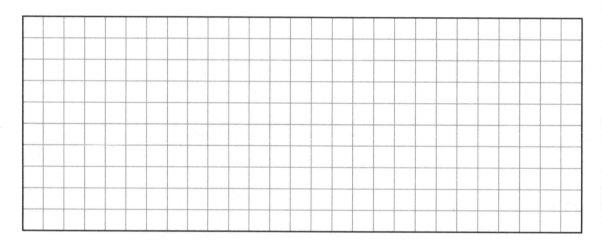

4 Count the triangles you can see on this shape.

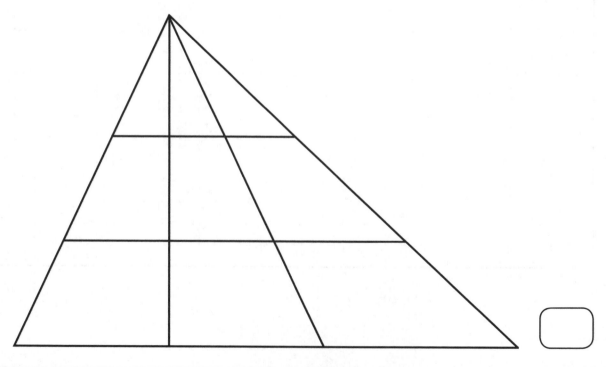

Reading scales and fractions

11a Millilitres

1 How many millilitres are in each measuring cylinder?

a

ml

c
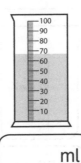

ml

e

ml

b

ml

d
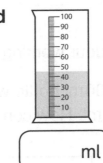

ml

f
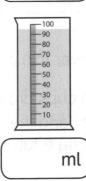

ml

2 Write the amount of water in each jug.

a
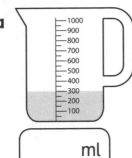

ml

c
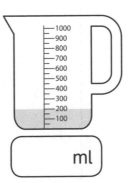

ml

e
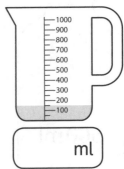

ml

b
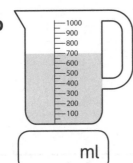

ml

d
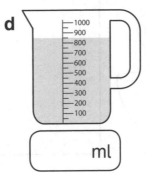

ml

f
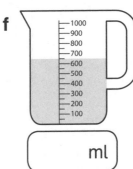

ml

Shade each jug to show the correct amount of water.

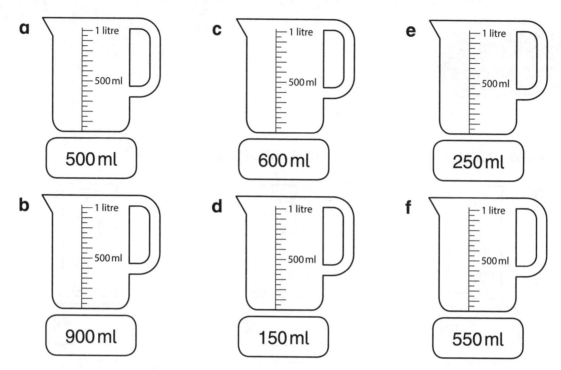

a 500 ml

b 900 ml

c 600 ml

d 150 ml

e 250 ml

f 550 ml

Answer this problem. Show your working in the box.

Two cups are used to fill a 500 ml bottle with water. One cup holds 40 ml and the other holds 70 ml. How can the bottle be filled with exactly 8 full cups?

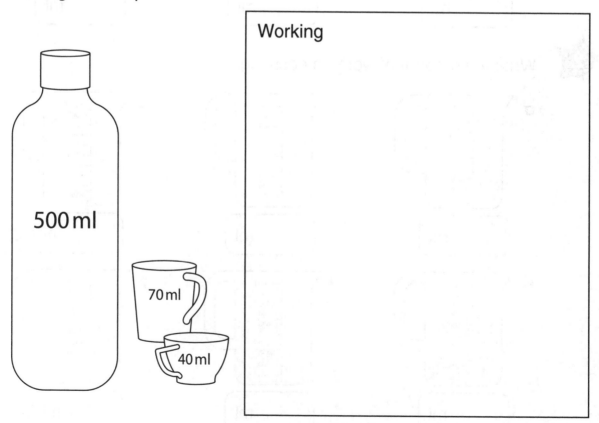

500 ml

70 ml

40 ml

Working

5

Try this estimating activity with a partner.

100 ml

Work with a partner and take turns.

- Choose 1 of the bottles.

- Fill the 100 ml container with water.

- Estimate where the 100 ml of water would come to on the bottle.
  Put an elastic band around the bottle at that point.

- Your partner then says whether they think that 100 ml is higher or
  lower than your estimate.

- Pour the water from the container into the bottle to find out.

- If your partner is correct, they score 2 points.

- If your partner is wrong, you score 2 points.

- Continue using different bottles and containers.

1 Complete these.

a 2 weeks = ☐ days

d 70 days = ☐ weeks

b 120 minutes = ☐ hours

e 3 years = ☐ months

c 3 hours = ☐ minutes

f $\frac{1}{2}$ hour = ☐ minutes

2 Draw the times on the clock faces.

a

3:20

c

2:50

e

1:55

b

6:45

d

10:05

f

6:35

3 Write the times shown on each clock face.

a

[:]

c

[:]

e

[:]

b

[:]

d

[:]

f

[:]

 4 How many minutes are there between these times?

a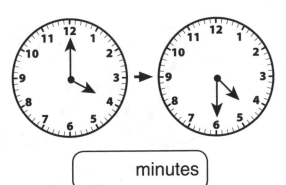

[minutes]

c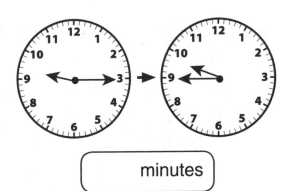

[minutes]

b

[minutes]

d

[minutes]

5 Look at these pairs of clocks.

How many minutes have passed between the times on these clocks?

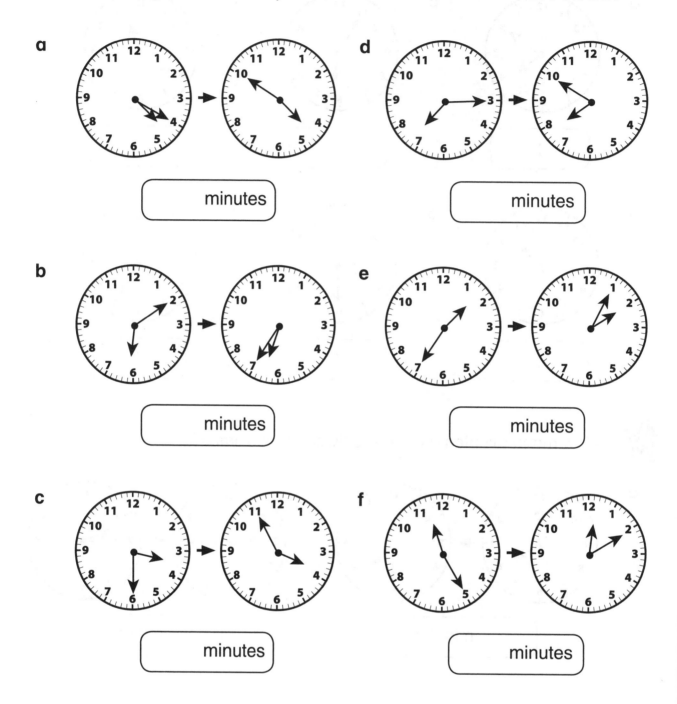

a

minutes

b

minutes

c

minutes

d

minutes

e

minutes

f

minutes

6

Read and answer these.

a Sam and his family leave home at 20 minutes past 8 in the morning and drive to school. They arrive at 9 o'clock. How long had they been driving?

> minutes

b A TV programme started at 5:15 and finished at 6:00. How long was the programme?

> minutes

c A pizza is put into the oven at 1:40 and takes half an hour to cook. What time must the pizza be taken out of the oven?

>

d It starts to rain at 10:35 and stops at 11:10. How long was it raining?

> minutes

e Jo starts a piano lesson at half past four. The lesson ends at quarter past 5. How long is her piano lesson?

> minutes

f A bus leaves the station at 9:55 and arrives at the market at 10:20. How long is the journey from the station to the market?

> minutes

g Gran's 90th birthday party is from 7:00 to 8:30. How long is the party?

>

h Danny pays to swim for 1 hour. He starts swimming at 1:15. What time must he finish swimming?

>

 1 What fraction is shown by the shaded part of each shape?

Choose from the list of fractions.

half	third	quarter

a

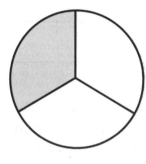

c

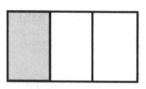

e

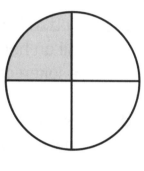

b

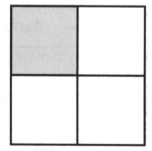

d

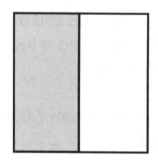

f

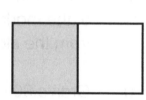

YOU WILL NEED:
• **coloured pencils**

Colour the shapes that show thirds ➜ 3 equal parts.

a
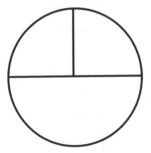

c
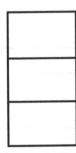

e
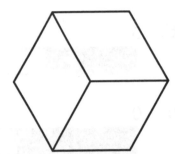

b
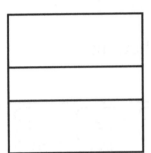

d
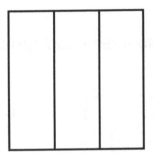

3 Draw lines to show where you would cut each cake.

YOU WILL NEED:
• **coloured pencils**

a Cut these into 2 equal parts. Colour $\frac{1}{2}$.

b Cut these into 3 equal parts. Colour $\frac{1}{3}$.
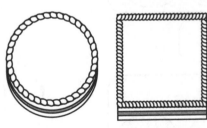 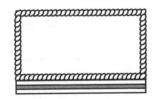

c Cut these into 4 equal parts. Colour $\frac{1}{4}$.
  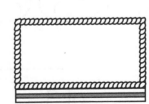

4   Write the fraction shown on each number track.

a   0 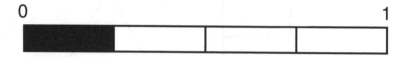 1   ☐

b   0  1   ☐

c   0 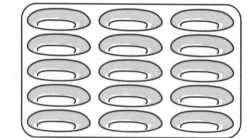 1   ☐

 5   Divide each tray into three equal groups. Count how many are in one-third.

a

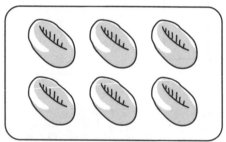

$\frac{1}{3}$ of 6 = ☐

d

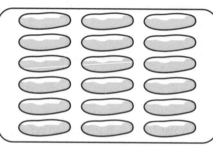

$\frac{1}{3}$ of 18 = ☐

b

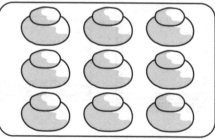

$\frac{1}{3}$ of 9 = ☐

e

$\frac{1}{3}$ of 15 = ☐

c

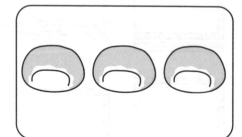

$\frac{1}{3}$ of 3 = ☐

f

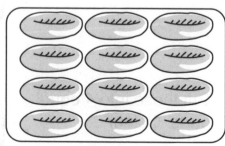

$\frac{1}{3}$ of 12 = ☐

Colour $\frac{1}{3}$ of each wall. Make each pattern different.

a

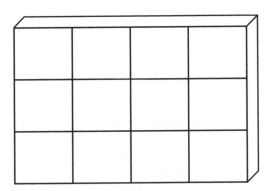

$\frac{1}{3}$ of 12 = ◯

c

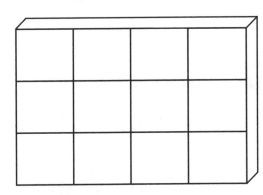

$\frac{1}{3}$ of 12 = ◯

b

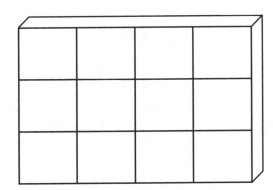

$\frac{1}{3}$ of 12 = ◯

d

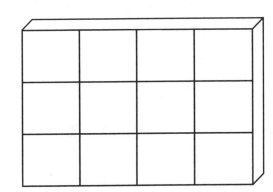

$\frac{1}{3}$ of 12 = ◯

12a Add or subtract?

1 Answer each set as quickly as you can.

a	b	c	d
6 + 9 =	50 + 70 =	14 – 9 =	130 – 70 =
4 + 14 =	20 + 90 =	17 – 6 =	90 – 50 =
7 + 8 =	40 + 80 =	16 – 8 =	160 – 90 =
11 + 5 =	80 + 60 =	12 – 9 =	140 – 80 =
9 + 8 =	30 + 70 =	19 – 12 =	150 – 60 =
6 + 13 =	60 + 120 =	16 – 7 =	170 – 110 =
7 + 7 =	110 + 40 =	17 – 9 =	110 – 90 =
5 + 6 =	50 + 130 =	15 – 6 =	150 – 120 =
8 + 8 =	90 + 80 =	18 – 14 =	190 – 40 =
7 + 9 =	120 + 70 =	13 – 11 =	150 – 70 =

2 Answer these. Complete the bar diagrams to help you.

a A lorry driver travelled 58 km in the morning and 40 km in the afternoon.
How far did the lorry travel in total?

b A stall had 60 bottles of apple juice at the start of a day and sold 48 bottles.
How many bottles were left?

c A postman had 99 letters and has delivered 39 letters. How many letters are left to deliver?

d A farmer has 35 chickens and 15 ducks. How many chickens and ducks are there altogether?

e Pip has read 76 pages of her book. She has another 80 pages left to read. How many pages in total are there in Pip's reading book?

Working

3 Answer these. Draw bar models to help you.

a Jan bought a pencil for 29p and paid with a 50p coin. How much change did Jan get?

b On a school trip there are two buses. One has 53 children and the other has 47 children. How many children in total are on the school trip?

c Kim is 8 years old. His gran is 75 years older than him. How old is Kim's gran?

d A coat cost £49 and shoes cost £32. What is the total cost of buying the shoes and coat?

4 Read and answer these. Show how you worked each of them out.

a What is 15 more than 78?

e How much greater is 91 than 76?

b What is the difference between 28 and 48?

f What is the total of 33 and 39?

c What number is 34 less than 52?

g What is 66 subtract 37?

d Add 57 and 26.

h Increase 124 by 47.

YOU WILL NEED:
• 5p and 2p coins

Use the bar diagram and some coins to help find the answer.

There are some 2p coins and 5p coins in a jar.

Altogether there is 53p. There are two more 2p coins than 5p coins.

How many of each coin are there in the jar?

Complete the chart to help you.

53p	

Number of coins	2p	5p
1	2p	5p
2	4p	10p
3		
4		
5		
6		
7		
8		
9		
10		
11		
12		

1 Look at the number machines.

Complete each table.

IN   OUT

+25

a

IN	12		32		52	
OUT		39		59		79

IN   OUT

−14

b

IN	38		48		58	
OUT		25		35		45

2 Join pairs that total 42. Check by adding in a different order.

9   28   26   25

17   14   33   16

3 Write a checking number statement for each calculation.

a 54 + 32 = 86

d 77 − 32 = 45

b 19 + 26 = 45

e 65 − 19 = 46

c 48 + 43 = 91

f 51 − 34 = 17

4 Answer these. Check each answer.

A	B	C	D	E
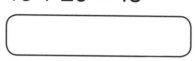				
48p	30p	27p	20p	34p

a How much greater in price is card A than card B? ⬚ p

b What is the difference in price between cards C and E? ⬚ p

c Which card is 28p less than card A? ⬚

d Which 2 cards have a difference in price of 4p? ⬚

e 2 cards cost a total of 82p. The difference in price between them is 14p. What is the cost of each card? ⬚ p ⬚ p

5 Subtract these and complete the number puzzle. Check your answers with an addition.

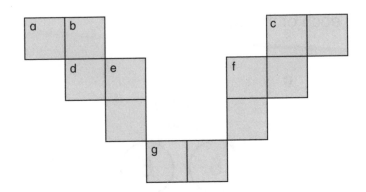

Across

a   78 − 62 = ⬭

c   86 − 53 = ⬭

d   49 − 25 = ⬭

f   99 − 74 = ⬭

g   68 − 51 = ⬭

Down

b   94 − 32 = ⬭

c   78 − 43 = ⬭

e   59 − 14 = ⬭

f   98 − 76 = ⬭

1 Write the 4 maths facts for each of these.

a

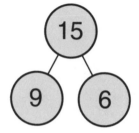

$\boxed{} + 6 = 15$

$6 + \boxed{} = \boxed{}$

$15 - \boxed{} = 6$

$\boxed{} - 6 = \boxed{}$

c

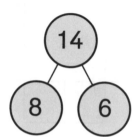

$\boxed{} + \boxed{} = \boxed{}$

$\boxed{} + \boxed{} = \boxed{}$

$\boxed{} - \boxed{} = \boxed{}$

$\boxed{} - \boxed{} = \boxed{}$

b

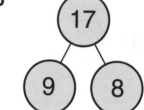

$9 + \boxed{} = 17$

$\boxed{} + 9 = \boxed{}$

$17 - \boxed{} = 9$

$\boxed{} - 9 = \boxed{}$

d

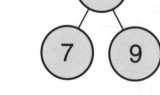

$\boxed{} + \boxed{} = \boxed{}$

$\boxed{} + \boxed{} = \boxed{}$

$\boxed{} - \boxed{} = \boxed{}$

$\boxed{} - \boxed{} = \boxed{}$

2 Use each of the numbers 5, 6, 7, 8, 9 and 10 to fill in the six missing numbers.

Only use each number once.

⬚ + ⬚ = 15   ⬚ + ⬚ > 15   ⬚ + ⬚ < 15

Can you find different ways to do this?

⬚ + ⬚ = 15   ⬚ + ⬚ > 15   ⬚ + ⬚ < 15

⬚ + ⬚ = 15   ⬚ + ⬚ > 15   ⬚ + ⬚ < 15

3 Write the missing numbers.

a $15 + \boxed{} = 38$

c $74 - \boxed{} = 21$

b $\boxed{} + 36 = 84$

d $\boxed{} - 59 = 35$

4 Write the missing numbers on these addition grids.

a

+	33	54
6	39	
7		

b

+	65	46
9		55
8		

5 Complete these addition walls.

a

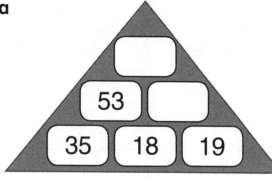

c

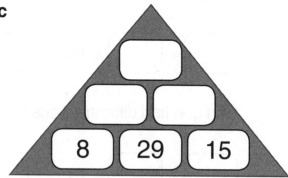

b

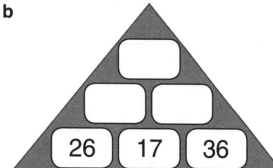

d

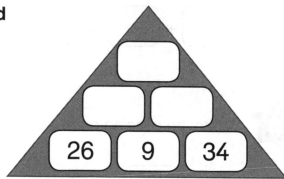

6

a Look at this calculation:

 $\boxed{}\ 5 + 8 = \boxed{}\boxed{}$

Write a digit in each box so that the calculation is correct.

Which other ways can you do it?

What patterns do you notice?

b Look at this calculation:

$\boxed{}\ 2 - 7 = \boxed{}\boxed{}$

Write a digit in each box so that the calculation is correct.

Which other ways can you do it?

What patterns do you notice?

1 Complete these.

a   20 + 8   28
   +   5   +   5
   _____   →   ___
      +
   _____   ___

d   40 + 4   44
   +   9   +   9
   _____   →   ___
      +
   _____   ___

b   70 + 5   75
   +   6   +   6
   _____   →   ___
      +
   _____   ___

e   90 + 3   93
   +   7   +   7
   _____   →   ___
      +
   _____   ___

c   40 + 6   46
   +   4   +   4
   _____   →   ___
      +
   _____   ___

f   80 + 7   87
   +   8   +   8
   _____   →   ___
      +
   _____   ___

2 Write the missing digits in each addition.

a
$$
\begin{array}{r}
4\bigcirc \\
+\ 2\ 4 \\
\hline
6\ 7
\end{array}
$$

d
$$
\begin{array}{r}
3\ 7 \\
+\ 3\bigcirc \\
\hline
7\ 2
\end{array}
$$

b
$$
\begin{array}{r}
4\ 8 \\
+\ \bigcirc 1 \\
\hline
7\ 9
\end{array}
$$

e
$$
\begin{array}{r}
\bigcirc 8 \\
+\ 2\ 8 \\
\hline
9\ 6
\end{array}
$$

c
$$
\begin{array}{r}
2\bigcirc \\
+\ 5\ 8 \\
\hline
8\ 4
\end{array}
$$

f
$$
\begin{array}{r}
1\ 8 \\
+\ \bigcirc 6 \\
\hline
3\ 4
\end{array}
$$

 3 The digits 1 to 8 are missing from these additions.

Put the digits in the correct places to complete them.

$$
\begin{array}{r}
\square\ 7\ 6 \\
+\ \ \ \square\ 2 \\
\hline
7\ 0\ \square
\end{array}
\qquad
\begin{array}{r}
8\ \square \\
+\ \square\ 9 \\
\hline
1\ 3\ 1
\end{array}
\qquad
\begin{array}{r}
9\ \square \\
+\ \square\ 4 \\
\hline
\square\ 5\ 1
\end{array}
$$

YOU WILL NEED:
• **digit cards 2–5**

Use the digits 2–5.

Place the digits in these empty boxes.

• Make the 2 answers as close together as possible.

• What if the answers have to be as far apart as possible?

Counting in threes, fractions and time

13a Multiplication table for 3

1 Answer these.

a 3 × 1 = ☐ ⋮
   1 × 3 = ☐

b 3 × 2 = ☐ ⋮⋮
   2 × 3 = ☐

c 3 × 4 = ☐ ⋮⋮⋮⋮
   4 × 3 = ☐

d 3 × 8 = ☐
   8 × 3 = ☐

e 3 × 10 = ☐
   10 × 3 = ☐

f 3 × 5 = ☐
   5 × 3 = ☐

g 3 × 3 = ☐ ⋮⋮⋮

h 3 × 6 = ☐
   6 × 3 = ☐

i 3 × 9 = ☐
   9 × 3 = ☐

j 3 × 12 = ☐
   12 × 3 = ☐

k 3 × 11 = ☐
   11 × 3 = ☐

l 3 × 7 = ☐
   7 × 3 = ☐

2 Draw jumps of 3 on this number line. Write the numbers you land on.

a

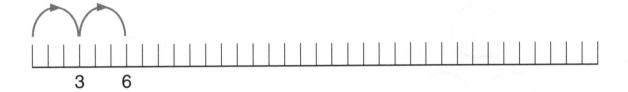

3   6

 3 Complete these. Learn these facts.

a   3 × 1 = ⬭   g   3 × 7 = ⬭

b   3 × 2 = ⬭   h   3 × 8 = ⬭

c   3 × 3 = ⬭   i   3 × 9 = ⬭

d   3 × 4 = ⬭   j   3 × 10 = ⬭

e   3 × 5 = ⬭   k   3 × 11 = ⬭

f   3 × 6 = ⬭   l   3 × 12 = ⬭

Fill in the missing numbers.

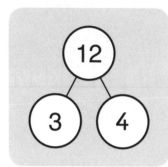

a

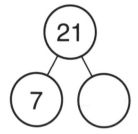

d

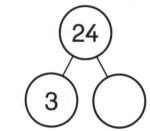

b

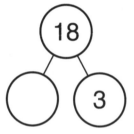

e

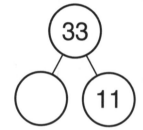

c

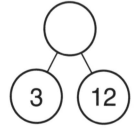

f

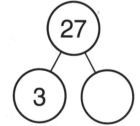

Look at these number machines and complete each table of results.

IN   OUT

a

IN	3	6	4	8	12
OUT					

IN   OUT

b

IN					
OUT	15	30	33	27	21

 5

Circle the numbers in the 3 times table.

26	2	27	4	19	30	8	14
	5	1	12	24	15	29	3
17	6	20	25	11	28	7	10
	13	23	16	22	21	9	18

1 Some of these shapes are divided into equal parts, some are not in equal parts. Circle the correct statement for each of these.

a

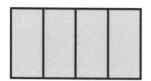

quarters

not quarters

c

thirds

not thirds

e

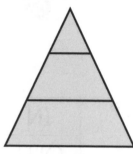

thirds

not thirds

b

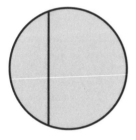

halves

not halves

d

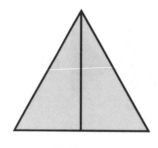

halves

not halves

f

quarters

not quarters

2 What fraction does the shading show in each shape?

Choose from this list of fractions.

$\frac{1}{2}$ $\frac{1}{3}$ $\frac{1}{4}$ $\frac{2}{3}$ $\frac{3}{4}$

a

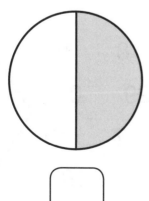

b

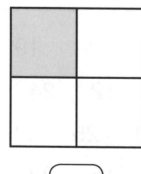

c

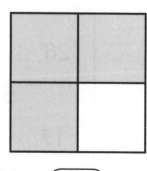

d   **e**

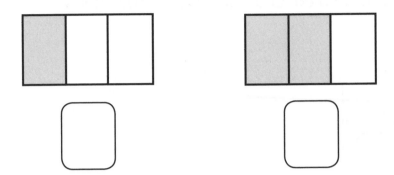

 3   Draw lines to show where you would cut each cake.

a   Cut these into halves.

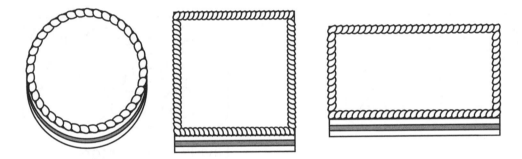

b   Cut these into thirds.

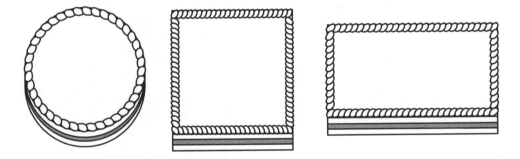

c   Cut these into quarters.

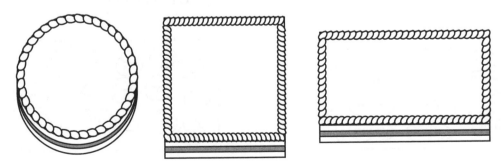

Divide each plate of rolls into equal groups. Count how many are in one group.

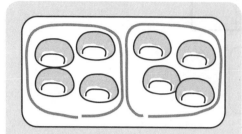

Divide into halves.

$\frac{1}{2}$ of 8 = 4

a

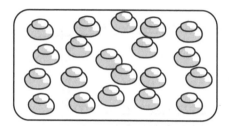

Divide into quarters.

$\frac{1}{4}$ of 20 =

d

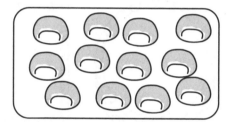

Divide into quarters.

$\frac{1}{4}$ of 12 =

b

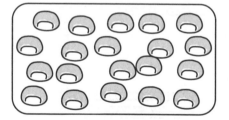

Divide into halves.

$\frac{1}{2}$ of 20 =

e

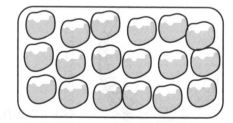

Divide into thirds.

$\frac{1}{3}$ of 18 =

c

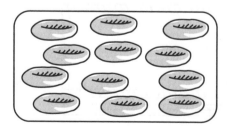

Divide into thirds.

$\frac{1}{3}$ of 12 =

f

Divide into halves.

$\frac{1}{2}$ of 18 =

5

Design your own flag so that $\frac{1}{4}$ of it is red and $\frac{1}{2}$ of it is yellow.

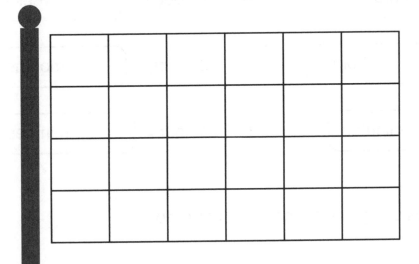

a What fraction of the flag is left white? ⬚

b How many squares are red? $\frac{1}{4}$ of 24 = ⬚

c How many squares are yellow? $\frac{1}{2}$ of 24 = ⬚

d How many squares are white? ⬚ of 24 = ⬚

1 Complete these.

a 1 day = [hours]

d 21 days = [weeks]

b 180 minutes = [hours]

e 2 years = [months]

c 2 hours = [minutes]

f $\frac{1}{4}$ hour = [minutes]

2 Write these times using quarter past or quarter to.

a

[   ]

d

[   ]

b

[   ]

e

[   ]

c

[   ]

f

[   ]

g

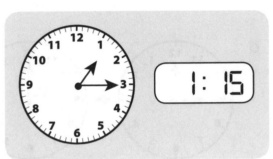

h

3 Write these times as they would be seen on a digital clock.

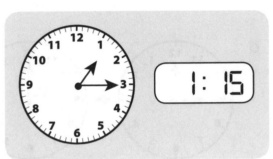

`1: 15`

a

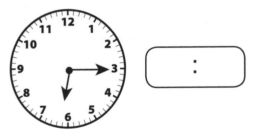

`:`

d

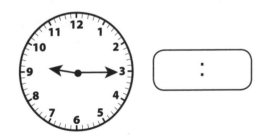

`:`

b

`:`

e

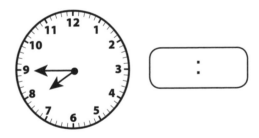

`:`

c

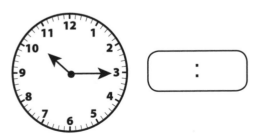

`:`

f

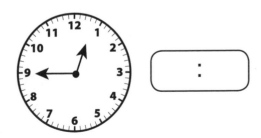

`:`

4 How many minutes are there between these times?

a

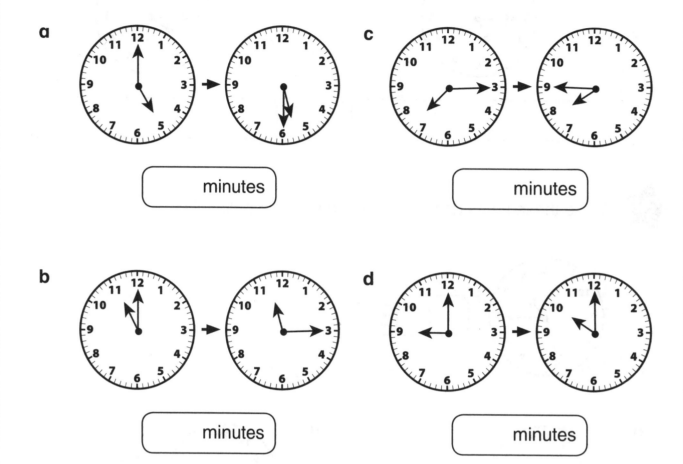

| minutes |

c

| minutes |

b

| minutes |

d

| minutes |

5 Draw the later time for each clock.

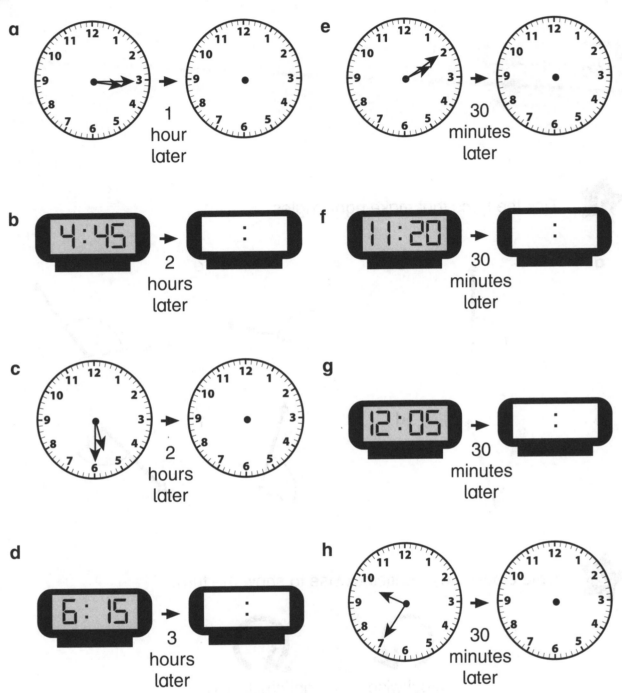

a   1 hour later

e   30 minutes later

b   4:45   2 hours later

f   11:20   30 minutes later

c   2 hours later

g   12:05   30 minutes later

d   6:15   3 hours later

h   30 minutes later

14a Turns

1 Tick the turns that make right angles.

a

c

e

b

d

f

2 Write clockwise or anticlockwise to show the turn.

 clockwise  anticlockwise

a open this tap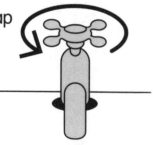

b open this door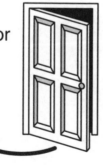

c   take off a bottle cap

d   turn the sound up

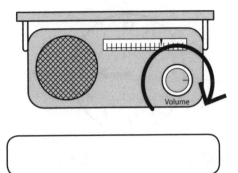

3   The hour hand makes the following turns. What number does it stop on?

a

quarter turn clockwise ☐

c

three-quarter turn clockwise ☐

b

half a turn clockwise ☐

d

whole turn clockwise ☐

The hour hand makes the following turns. What number does it stop on?

a

half a turn clockwise ☐

c

three-quarter turn clockwise ☐

b

quarter turn clockwise ☐

d

half a turn clockwise ☐

1 Join the matching pairs.

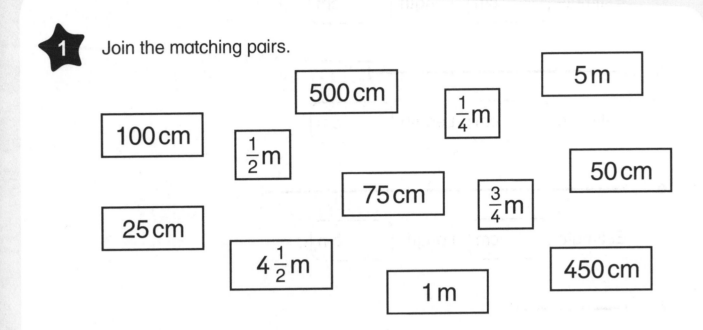

500 cm

5 m

$\frac{1}{4}$ m

100 cm

$\frac{1}{2}$ m

50 cm

75 cm

$\frac{3}{4}$ m

25 cm

4 $\frac{1}{2}$ m

450 cm

1 m

2 YOU WILL NEED:
• a ruler

Look at this ruler. Estimate the length of each line.

Use a ruler to measure the exact length of each line.

Write your estimate and the exact length.

a ├────────────┤

Estimate [cm] Length [cm]

b ├──────────────────┤

Estimate [cm] Length [cm]

c

Estimate [cm] Length [cm]

d

Estimate [cm] Length [cm]

e

Estimate [cm] Length [cm]

f

Estimate [cm] Length [cm]

3 Estimate and measure the length of each line.

Think about how you will measure them.

YOU WILL NEED:
• a ruler

a

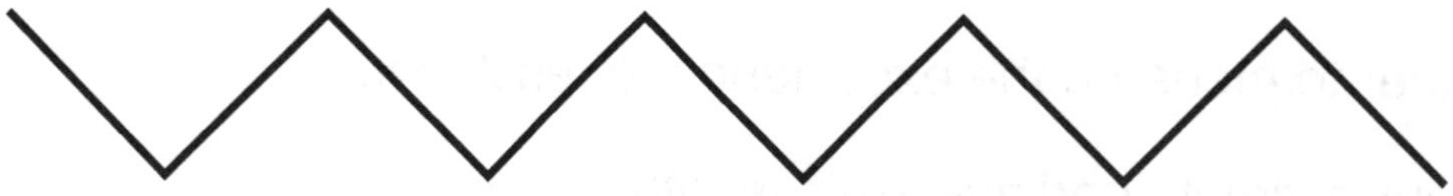

Estimate [cm] Length [cm]

b

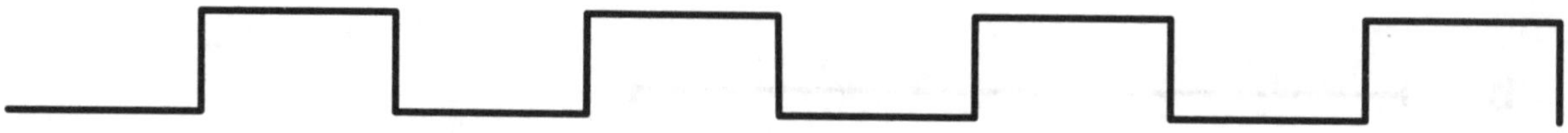

Estimate [cm] Length [cm]

4

Estimate and measure these using a ruler or metre stick.

Then choose 5 more things to measure to complete the chart.

I measured	I estimated	It was
distance from your desk to the door		
height of your desk		
length of a pencil		

YOU WILL NEED:
- a metre stick
- marbles
- beanbags
- two books

Measure and mark out a 1 m track on the floor.

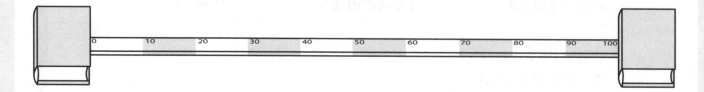

a Try to roll a marble exactly 1 metre.

b Try to throw a beanbag exactly 1 metre.

c Turn away from the track. Without measuring, place two books so that they will be 1 metre apart. Measure the distance. Is it shorter or longer than 1 metre?

⭐ 1   Complete this chart to show where Josh is facing after each turn.

home

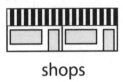

shops

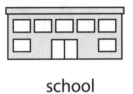

school

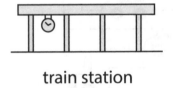

train station

Start position, facing	Turn	End position, now facing
home	$\frac{1}{4}$ turn anticlockwise	
train station	$\frac{1}{2}$ turn clockwise	
school	$\frac{1}{4}$ turn clockwise	
shops	$\frac{1}{4}$ turn anticlockwise	

2

YOU WILL NEED:
• colouring pencils

Colour each route red, blue or green.

Count the right angles on each route.

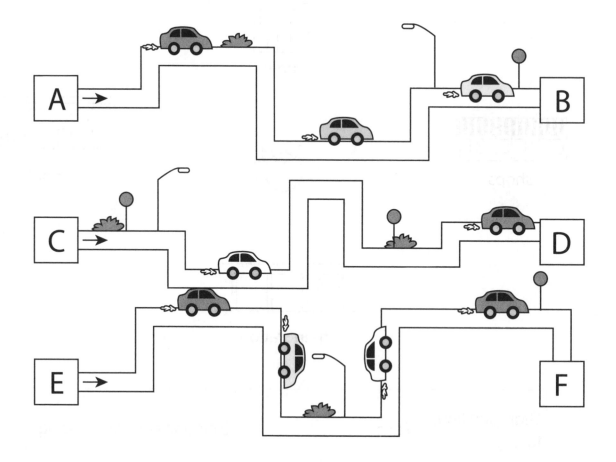

Blue route   A ➡ B   (right angles)

Red route   C ➡ D   (right angles)

Green route   E ➡ F   (right angles)

3 Follow these directions. Where do the spaceships each go?

a SPACESHIP A
Go forward 2.
Turn 1 right angle clockwise.
Go forward 2.
Turn 1 right angle anticlockwise.
Go forward 3.
Turn 1 right angle clockwise.
Go forward 2.
Turn 1 right angle clockwise.
Go forward 5.

FINISH:

b SPACESHIP B
Go forward 3.
Turn 1 right angle anticlockwise.
Go forward 1.
Turn 1 right angle clockwise.
Go forward 2
Turn 1 right angle anticlockwise.
Go forward 3.
Turn 1 right angle clockwise.
Go forward 2.
Turn 1 right angle clockwise.
Go forward 4.

FINISH:

c SPACESHIP C
Go forward 1.
Turn 1 right angle clockwise.
Go forward 5.
Turn 1 right angle clockwise.
Go forward 2.
Turn 1 right angle clockwise.
Go forward 1.
Turn 1 right angle anticlockwise.
Go forward 1.
Turn 1 right angle clockwise.
Go forward 2.
Turn 1 right angle clockwise.
Go forward 1.
Turn 1 right angle anticlockwise.
Go forward 1.

FINISH:

d Make up your own routes for the spaceships on the map.

4 Complete these instructions to move the robot from start to finish.

Turn the robot to face the correct direction at the start.

START			
			FINISH

a **Two steps**

Turn ⬜ right angle ⬜⬜⬜⬜⬜⬜ Go forward ⬜

Turn ⬜ right angle ⬜⬜⬜⬜⬜⬜ Go forward ⬜

b **Four steps**

Turn ⬜ right angle ⬜⬜⬜⬜⬜⬜ Go forward ⬜

Turn ⬜ right angle ⬜⬜⬜⬜⬜⬜ Go forward ⬜

Turn ⬜ right angle ⬜⬜⬜⬜⬜⬜ Go forward ⬜

Turn ⬜ right angle ⬜⬜⬜⬜⬜⬜ Go forward ⬜

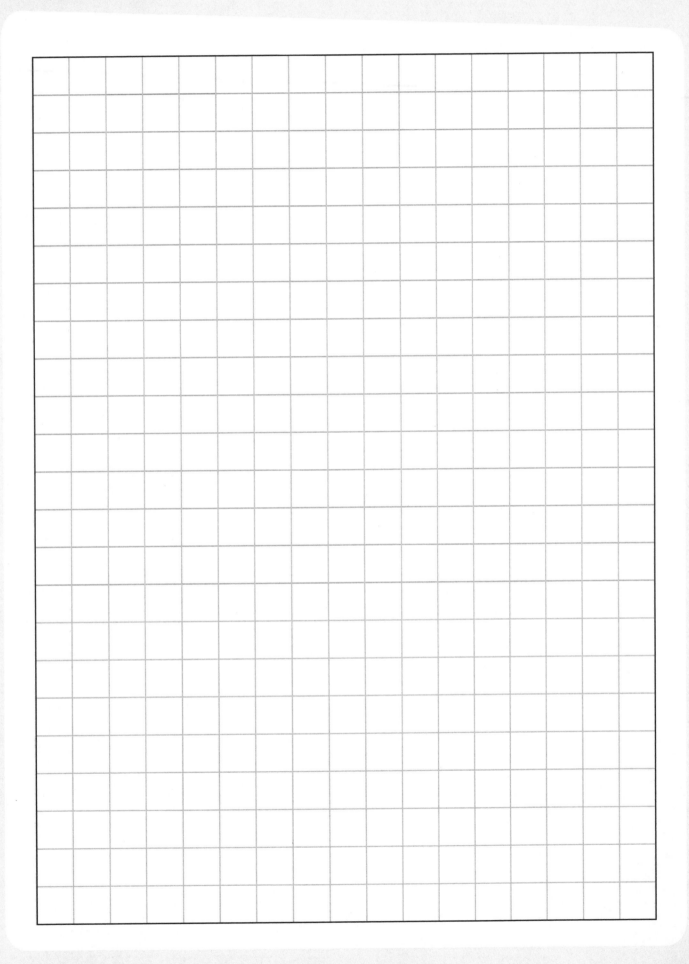

RISING STARS Mathematics

Rising Stars Mathematics includes high-quality Textbooks, Teacher's Guides, Practice Books, online tools and CPD to provide a comprehensive mastery programme for mathematics.

For more information on the complete range, visit
www.risingstars-uk.com/rsmathematics.

RISING STARS Mathematics

Practise all the skills you have learnt in class with Seb and Lili.

Year 2c

ISBN 978-1-78339-815-7

9 781783 398157 >

RISING STARS

For more information please call 01235 400 555

www.risingstars-uk.com

Follows the NCETM textbook guidance